T0225203

CAMBRIDGE UNIVERSITY PRESS
Cambridge, New York, Melbourne, Madrid, Cape Town, Singapore, São Paulo, Delhi

Cambridge University Press
The Edinburgh Building, Cambridge CB2 8RU, UK

Published in the United States of America by Cambridge University Press, New York

www.cambridge.org
Information on this title: www.cambridge.org/9780521204101

© Cambridge University Press 1973

First published 1973
Re-issued in this digitally printed version 2008

A catalogue record for this publication is available from the British Library

ISBN 978-0-521-20410-1 paperback

Everyone in this University seems to know that the Jacksonian Chair was founded for the study of gout, but there are other and more substantial reasons why I, the latest holder of the Chair, should feel somewhat overawed by the consequences of the Founder's will. The Chair was established in 1783, the last to be founded in the eighteenth century, by the will of the Reverend Richard Jackson, M.A., a Fellow of Trinity, who enjoined that the Professor was 'to be chosen for his knowledge in Natural Experimental Philosophy and of Chymistry'. He was to be allowed a wide latitude in his lectures, provided that they were of an experimental nature, but natural bodies which were the subject of enquiry were to be exhibited at lectures. He was requested 'to have an eye more particularly to that *opprobrium medicorum* called the gout'. The General Board, by assigning the Chair to the Faculty of Physics and Chemistry, seems to have recognised that most holders have had no competence (or experience) in the study of gout, but the Electors have hitherto been indeed successful in carrying out the substantial intention of Jackson's will, for most of the Professors have been notable exponents of experimental philosophy.

The first holder was Isaac Milner of Queens', who later became Lucasian Professor in 1798. He was followed by F. J. M. Wollaston (Trinity Hall), the brother of the metallurgist and mineralogist, William Hyde Wollaston, whom we shall meet later, and he in 1813 by W. Farish (Magdalene), who had previously been Professor of Chemistry. The fourth Professor was Robert Willis (Caius), more notable, I believe, for his architecture than for experimental philosophy; he was Jacksonian Professor for thirty-eight years. Professor Charlton (1972) has recently remarked that three of the first four Jacksonian Professors, Milner, Farish and Willis, were among the first in this country to lecture on the application of science to engineering. Of these three, Willis was the most notable; he worked on the theory of mechanisms (Willis, 1841) and served on the Government committee to study the behaviour of railway bridges under moving traffic. But chiefly he was a learned architect (Pevsner, 1970); his principal achievements were in architectural history and archaeology, especially of Canterbury Cathedral, but he was also a practitioner – he designed the west window of St Botolph's and, in collaboration with Whewell, the wooden vault of the Great Gate of Trinity. His greatest monument is his magnificent account of the buildings of Cambridge and Eton – 'Willis and Clark' (Willis and Clark, 1886).

Willis was succeeded in 1875 by James Dewar (Peterhouse). Dewar's work in physics and chemistry was extensive, and he is the first holder of the Chair to have made a major contribution to experimental physics as we practise it today. Together with Liveing (Liveing and Dewar 1915), he conducted extensive studies in spectroscopy, lasting for many years, and for all but two years of his incumbency of the Jacksonian Chair he was also Director of the Royal Institution, where he carried out his work on the liquefaction of the 'permanent' gases. Dewar's first studies in spectroscopy may seem far from our present aims and methods, done as they were with cumbersome apparatus and without the guidance of quantum mechanics, but he so developed the subject that his later work has a modern feel; as for his work on liquid gases, try to imagine physics or engineering or daily life without the vacuum flask that he invented. Dewar's incumbency was fifty years and, with the present retiring age of sixty-seven, it is unlikely ever to be exceeded. He was succeeded by C. T. R. Wilson, who was Professor until 1936– again try to imagine nuclear and high-energy physics and meteorology without Wilson's work on the cloud chamber. Sir Edward Appleton, who established the existence of the ionosphere, succeeded Wilson in 1936, and Sir John Cockcroft, who built the first particle accelerator for nuclear physics, followed him in 1939 when Appleton went to the Department of Scientific and Industrial Research (D.S.I.R.), as it then was; Professor Frisch, who had helped to discover the fission of nuclei, was elected in 1947. Surely Richard Jackson must be well satisfied with the successive choices the Electors have hitherto made of Professors 'chosen for [their] knowledge in Natural Experimental Philosophy'.

I am indeed at once overawed and deeply gratified at being elected in such a succession and, coming as I have from the University of Edinburgh, I feel particular satisfaction in occupying this Chair. Sir James Dewar was an Edinburgh graduate and C. T. R. Wilson was born of a farming family of Glencorse, just south of Edinbrugh, retiring to Carlops in Peebleshire, where his daughters still live. Wilson's interest in the physics of clouds began when he was an observer at the now defunct observatory on top of Ben Nevis. Sir Edward Appleton, after notable years at D.S.I.R., was appointed Principal of Edinburgh and initiated an interest in geophysics which fructified in the Chair I held there. It was a great pleasure for me to be asked last year to give the annual lecture established in his memory. Finally, the title of Natural Philosophy carries with it the humane feeling of the Scottish Enlightenment which made Edinburgh the Athens of the North in the eighteenth century, and whose spirit is still potent there.

THE NATURAL PHILOSOPHER

Let me enlarge a little on the idea of Natural Philosophy. As men and women we find ourselves able to observe and experience the natural world in a somewhat detached way, to marvel at what we see and to develop a rational and consistent account of the structure and development of at any rate parts of that world. These faculties, like the practice of the arts, mark us off from the brute creation and, like the arts, give those of us fortunate enough to be able to practise them, immense if sometimes transitory satisfaction. To develop the understanding and analysis of nature is a mark of a civilised community and to make them more widely appreciated must surely be an aim of civilised government, irrespective of the economical benefits of experimental science.

The rational understanding of nature is thus the aim of the natural philosopher. More particularly, the natural philosopher is one who follows the methods that are exemplified in the *Principia Philosophiae Naturalis* of Newton, although they were practised and understood much earlier, by William Gilbert of St John's in his study of the magnetisation of the Earth and by Galileo Galilei in his investigations in optics, mechanics and astronomy.

In essence, the natural philosopher *observes* nature, he abstracts the features of his observations that he considers essential and constructs hypotheses to reproduce them, and he *predicts* new phenomena and checks them against observation. It is the third stage that is characteristic of modern science and it is the predictive power of modern physics which gives us confidence that its theories do indeed correspond to the realities of nature independent of our subjective limitations and importations.

The strength of modern science is that its conclusions can be so stated that it is formally possible to prove them wrong. If a theory, a model that is, of the natural world fails of disproof in successive tests, one may have considerable confidence that it is indeed possible to give a rational account of the natural order and that the particular theory is a good description of how some part of the world is put together and behaves. Thus, the Michelson–Morley experiment gave a strong indication that a theory such as special relativity was required (there seems to be some argument about whether Einstein was aware of the experiment at the time he was developing special relativity) but it has nothing obviously to do with variations of mass with velocity, nor with the so-called spin of the electron. However, the consequences of special relativity include, on the one hand, Dirac's formulation of the relativistic wave equation for the electron, which demonstrated that electron spin was a consequence of relativistic invariance and so related all the data of atomic spectra to special relativity, and, on the other hand, the variation of mass with velocity that must be taken into account in the design of large particle accelerators. These two consequences, one the reduction to order of a wide range of atomic and molecular physics, the other a principle of design in heavy engineering, are proofs of the validity of special relativity of quite different strength from the original considerations which suggested its foundation. Thus, the predictive power of natural philosophy leads on the one hand to *engineering* – the understanding of nature that enables us to modify it – and on the other hand to ideas beyond the range of immediate day-to-day experience. We find ourselves able to extrapolate the behaviour of the natural world into conditions of which we so far have no direct knowledge. However, we all know that extrapolation is a most uncertain activity, for within a *limited* range of conditions a *wide* range of hypotheses may fit our experiences, and the consequences of those hypotheses in different conditions may be very diverse. If we are to extend and deepen our understanding of the natural world, we must endeavour to extend the range of conditions throughout which we test our hypotheses

A simple instance may perhaps help to make my point. Suppose we observe a variable y as a function of a variable x over a range δx, and suppose that the scatter of our observations is of the order of σ. Let the observed range of y be δy (Figure 1). Then our first shot at a functional relation between y and x will be

$$y = y_0 + \frac{\delta y}{\delta x}\,(x - x_0).$$

$\delta y/\delta x$ will be uncertain by about $\sigma/\delta x$, and, if we wish to estimate y when x is $K\delta x$ outside the original range, we shall find that the uncertainty of our prediction is $K\sigma$. However, this uncertainty depends on the linear relation between y and x being correct.

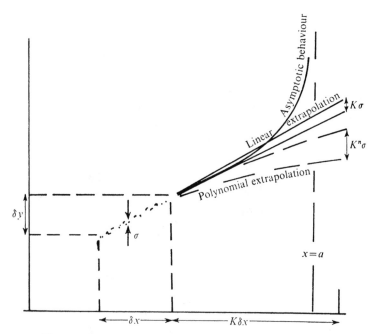

Figure 1. Extrapolation from a limited range of observations.
The conclusions that are drawn depend on the assumptions that are made and not solely upon the data.

Suppose in fact the correct relation is

$$y = y_0 + m(x - x_0) + f(x),$$

where $f(x)$ is some function that is not linear. All we know about $f(x)$ is its possible variation over the range δx, namely

$$f_{max} - f_{min} < 2\sigma,$$

8

where f_{max} and f_{min} are the maximum and minimum values of f in the range δx. Suppose $f(x)$ is $a(x - x_0)^n$. Then it easily follows that the estimate of a must be less than about $2\sigma/(\delta x)^n$, but when extrapolating over a range $K\delta x$, the $f(x)$ term may contribute up to $2\sigma K^n$. This shows that unless we know *a priori* the functional form of y, it is not really possible to estimate the uncertainty of our extrapolation.

So far, we have supposed that all regions in the x and y plane are accessible, but that is not necessarily so. Suppose y is actually a branch of an hyperbola with an asymptote at $x = a$, so that y approaches infinity as x approaches a. If we do not know that such an asymptote exists, we may predict the behaviour of y on the basis of a polynomial fitted to y within the range δx and we may conclude that there are values of y corresponding to values of x greater than a. In fact, that whole range of the plane is excluded; similarly, y may approach b as an asymptote and can never exceed it, although a polynomial extrapolation would predict larger values.

This simple example I take to be a model of how we set about understanding the natural world. When we cannot travel to Africa, it is possible to guess that men there have a large foot to use as a parasol. When we can do experiments only within the range of day-today experience, our knowledge of the world is very limited and our range of prediction about how it behaves can be very wide, so wide that not only may our predictions be quantitatively wrong, in that we get the general nature of the behaviour right but the orders of magnitude wrong, but they may be qualitatively wrong in that we predict phenomena in ranges of conditions where they do not occur.

Historically, astronomy has often afforded the natural philosopher the means of breaking the confines of his laboratory, and I believe it to be at the very least as important for the physicist to look to astronomy now as it has been in the past. I could illustrate my case by referring to William Gilbert, or Galileo, or Newton, or Halley, and to geophysics as part of astronomy, but I prefer to consider a more recent group of examples which lead up to current problems of physics and astronomy still of the greatest interest.

SOME HISTORICAL EXAMPLES

Fraunhofer lines and spectroscopic analysis

In early spectroscopic studies with the prism the light had been placed behind a pinhole, just as Newton had first done, but at the end of the eighteenth century W. H. Wollaston, the noted chemist and mineralogist, and brother of the second Jacksonian Professor, replaced the pinhole with a slit, so improving the resolution considerably. In consequence, he was the first to see the dark lines that cross the spectrum of the Sun, although he did not appreciate their significance and thought they marked the boundaries of the colours of the spectrum. Subsequently Fraunhofer in 1815 made an atlas of some dark lines and it was then noticed that two of the dark lines coincided with the bright yellow lines characteristic of the spectrum of common salt in a flame. Kirchhoff (1862, 1863) followed by constructing a very detailed map of the dark lines in the spectrum of the Sun (Plate 1) and by his fundamental study of the emission and absorption of light, as a result of which he formulated the rule that for any body the ratio of emissivity to absorptivity at the same wavelength depends only on temperature.

On the practical side, these ideas led, in the hands of Bunsen, to spectrochemical analysis. In fundamental physics, they contributed to the development of thermodynamics, while a deeper analysis at the hands of Einstein led to the concept of stimulated emission. I think it not too fanciful to trace a direct link between the laser and maser and Wollaston's innovation of the slit for the pinhole.

A number of distinct elements may be identified in this history. It began with a technical development which led to an astronomical discovery. That in turn led to laboratory and theoretical studies by which it was interpreted and, the interpretation being to hand, the astronomical observations could be used for the physical and chemical study of the atmosphere of stars. Subsequent theoretical analysis brought about fundamental improvements in the understanding of the interaction of radiation and matter. We shall find such elements and such a progression in other instances.

The Balmer series

The data originally available to Balmer when he began to study the regularities of wavelengths of lines in the spectrum of (atomic) hydrogen were the wavelengths of four lines in the visible, $H\alpha$, $H\beta$, $H\gamma$ and $H\delta$, measured by Ångström. To them he fitted the formula

$$\lambda = hm^2/(m^2 - n^2) \quad \text{(Balmer, 1885).}$$

He then heard that two observers, one of them Sir William Huggins, had already in 1880 measured lines in white stars (Huggins, 1880), the wavelengths extending into the ultra-violet (Plate 2). J. Johnstone Stoney was, I believe, the first to suggest, in a letter to Huggins (*loc. cit.*), that the hydrogen series followed a simple rule (Figure 2), but Balmer had the key and showed that Huggins's lines verified his formula. Balmer's work was the experimental basis of the early quantum theory of atomic spectra. It shows clearly one aspect of the relation between astronomy and natural philosophy – the possibility of extending the range of observation by making use of astronomical conditions, for at the time it was not possible to excite the higher levels of the hydrogen atom in the conditions that could be attained in laboratory sources. Huggins's observations were made

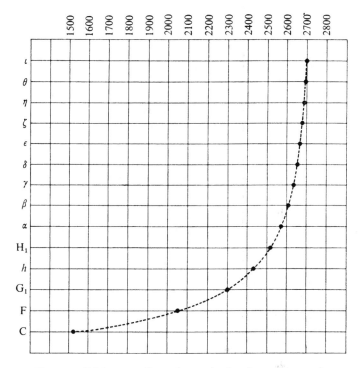

Figure 2. J. Johnstone Stoney's graph showing a systematic variation of the wavelengths of the ultra-violet lines of white stars (in Huggins, 1880).

possible by his application of dry-plate photography to astrophysics.

The discovery of helium

Janssen, in 1868, and Sir Norman Lockyer (1874) independently discovered a bright yellow line, called

D_3 because it was close to the D_1 and D_2 lines of sodium, in the spectrum of the chromosphere during an eclipse of the Sun (Plate 3). The line could be identified with none from known elements, and Lockyer suggested that it came from a new element which he proposed to call *helium*. Not until thirty years later was helium discovered on the Earth when, following the isolation of argon by Lord Rayleigh and Sir William Ramsey, Ramsey investigated the gas given off by the radioactive mineral cleveite and isolated a fraction giving the D_3 line in its spectrum.

In this instance, chemists had failed to discover helium on the Earth because the abundance of helium here is abnormally low. It makes up 10 per cent of the Sun and other normal cosmic objects, but because the gravitational pull of the Earth can retain only relatively dense atmospheric constituents, we have almost no elemental helium in the lower atmosphere and none in combination, for, unlike hydrogen, it is inert.

Forbidden lines

As my final historical example, I take the intense emission lines such as that at 3730 Å seen in the spectra of planetary nebula, the clouds of gas like the great nebula in Orion that sometimes surround very hot stars (Plate 4). The lines were at first attributed to an unknown element, e.g. *nebulium*, but with the completion of the periodic table of the elements it was

14

realised that there was no place for a new element, as there had been for helium, and accordingly the lines must come from some transition between energy levels of a known atom which had not been observed in the laboratory. The explanation was given by I. S. Bowen in 1928, who argued that the lines corresponded to so-called forbidden transitions in the spectra of singly and doubly ionised oxygen, O-II and O-III. Forbidden transitions are strictly transitions of very low probability and, if all the radiative transitions from a given state of an atom are of low probability, atoms would remain in that state for a long time if they could escape from it only by radiating. It is difficult to observe such transitions in laboratory sources because the density of gas is normally such that collisions with other atoms or electrons allow the atoms to escape from metastable states, as they are called. If the density is reduced so that collisions are infrequent, the total number of atoms is too low to give sufficient photons to observe radiation. Astronomical sources such as planetary nebulae and the solar corona are both of low density and of large total volume, so that while collisions are sufficiently infrequent to allow a reasonable population of atoms in metastable states to build up, at the same time the total number along the line of sight is large enough that even transitions of low probabilities can be observed.

Astronomy and the development of atomic spectroscopy

It is plain that from the early observations of Wollaston through to the identification of forbidden transitions, astronomical observations have been crucial in the development of atomic spectroscopy, and I believe that the reason is that astronomical sources afforded conditions which were not at the time accessible in laboratory sources. Fraunhofer lines were observed in the spectrum because the Sun's atmosphere provides a relatively cool absorbing layer of atoms in front of a hot radiating layer; the highly excited states of hydrogen in stars which provided Balmer with further instances of his rule for wavelengths of hydrogen lines could not then be produced in laboratory sources; helium is far more abundant in the Sun and stars than on Earth, and the combination of low density and large total number that is necessary for observing forbidden lines is difficult to produce in the laboratory.

Subsequently many similar conditions have been produced in the laboratory and more detailed controlled observations have been made than is possible on astronomical sources. The stimulus none the less came from astronomy.

There is one other point to make about my examples – in each case the astronomical observation and the physical development were the rather immediate outcome of a development in technique. I have mentioned Wollaston's replacement of a pinhole by a slit in the prism spectroscope; Huggins was able to measure the wavelengths of the higher members of the Balmer series because he had introduced dry photographic plates to record astronomical spectra; the discovery of helium in the Sun followed improvements in the method of observing the solar spectrum at a total eclipse, and it was the greater light-gathering power of large telescopes that allowed the study of the rather weak sources that planetary nebulae are.

RANGES OF CONDITIONS

At this point I should try to say what the conditions are that we may expect to establish in physics laboratories on the Earth, and to give some indication of how far outside those conditions astronomical conditions may take us, so that we may have an idea of the extent of the wider vistas that astronomy may afford us.

I shall take the properties of an isolated atom or molecule to establish a scale. The dissociation energy of a typical molecule is some 5 electron volts and the ionisation energy of a typical atom lies between 10 eV for neutral atoms and 1.5 keV for highly ionised atoms. The lifetime of an atom in a state that radiates through an allowed transition is of the order of 10^{-7}s or less. It is easy to produce plasmas with 50 V electrons in an electrical discharge in a gas, so that it is possible to excite most of the excited states of most neutral and many ionised atoms in that way. However, the temperature corresponding to 50 eV is nearly 0.5 MK, so that Boltzmann's law tells us that in laboratory sources with temperatures of a few thousand degrees there will be appreciable populations of atoms in only the lower excited levels, and only in high-temperature plasma devices such as ZETA can highly ionised atoms be produced. Fifty eV is the kinetic energy of protons travelling at 13 km/s or 4.10^{-5} of the velocity of light; it is not difficult to accelerate particles to such a speed in laboratory apparatus.

The interval between collisions of hydrogen atoms at a pressure of one-thousandth of an atmosphere is one-tenth of a microsecond, comparable with the radiative lifetime of an excited state. In consequence, the populations of excited states in most laboratory sources are determined by collisions, but the lifetimes with respect to collisions are long enough for many radiative transitions to occur. The main point about astronomical sources of atomic and molecular radiation is that the balance between radiative and collisional processes can be quite different. The densities can be much less than in laboratory sources yet the total number of photons emitted per unit area of source can be large because the number of atoms in a column along the line of sight can be large. Again, the rates of collision can be very low, allowing infrequent radiative processes

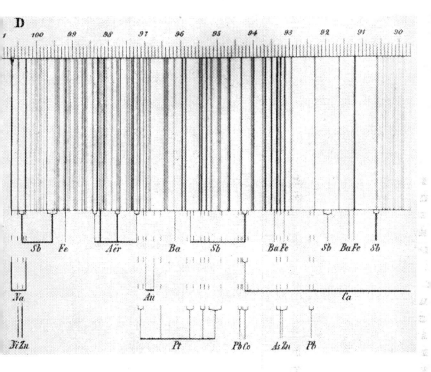

Plate 1. Portion of Kirchhoff's map of the solar spectrum, showing many dark lines (Kirchhoff, 1863).

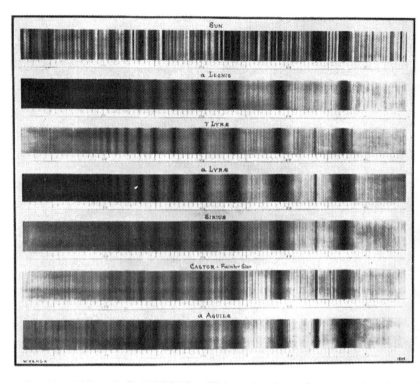

Plate 2. Photographs of ultra–violet lines of hydrogen in the spectra of white stars (Huggins, 1880).

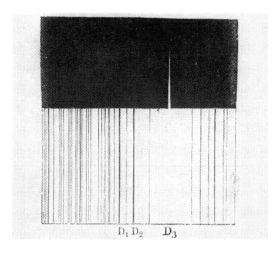

Plate 3. Drawing of the solar spectrum at eclipse showing the D_3 (yellow) line of helium (Lockyer, 1874).

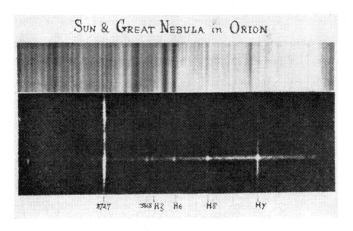

Plate 4. Forbidden line at 3730 Å in the spectrum of the great nebula in Orion, photographed by Sir William Huggins.

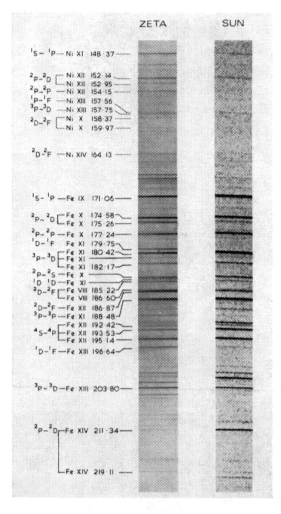

Plate 5. Comparison of part of the ultra-violet spectrum of the Sun's corona with the spectrum of the ZETA plasma (reproduced by permission of Dr A. H. Gabriel, Astrophysical Research Unit, Culham).

to occur, yet the lifetime of an astronomical source is usually sufficient for thermal equilibrium to be established. It is for such reasons that we can observe radiative transitions that are difficult to detect in terrestrial sources.

The elastic energy per atom of a solid at a pressure of 10^{11} N/m^2 (1 million atmospheres) is of the order of 10 eV, just about the ionisation energy of many atoms. At greater pressures we may expect that the behaviour of matter will hardly be affected by the details of the electronic structure of the atom. The pressure in the Earth attains some 3.10^{11} N/m^2 and that in Jupiter is probably ten times greater. A pressure greater than 10^{11} N/m^2 can be attained for a time of a microsecond when a strong shock wave passes through iron or aluminium or water, and to some extent, therefore, the properties of substances at pressures such that the elastic energy is comparable with the spacing of electronic energy levels can be studied experimentally. The pressure at the centre of Jupiter cannot yet be attained in shock waves, but Jupiter consists almost entirely of hydrogen and helium, and considerable progress has been made in calculating the properties of those elements at pressures of the order of 10^{12} N/m^2 – it is almost certain that both change to metallic forms. The stimulus for much of the theoretical work has come from astronomical studies of the mechanical properties of Jupiter and Saturn; much remains to be done and the properties of Uranus and Neptune are still a puzzle, but astronomy affords us samples of the behaviour of matter at much higher pressures still, in the white dwarfs and pulsars.

In summary, then, the conditions in which we may expect to see phenomena not predicted by our terrestrial laboratory experience are those in which energies of particles greatly exceed 100 eV and temperatures exceed 1 MK, in which pressures exceed 10^{11} N/m^2 or in which atomic events occur much less frequently than once every millisecond, or in which the effects of forces that, like gravity, are very small may be seen. I do not put this list forward as restrictive or exhaustive but as one which may give a general idea of the directions in which astronomical observations may be especially useful. It is likely indeed to be a somewhat ephemeral guide, for astronomical conditions are a standing challenge to experiments to reproduce them in the laboratory.

CURRENT PROBLEMS

A decade or so ago, I suspect that many physicists would have dismissed the history I have given you as indeed history, of little relevance to current physics, which was quite capable of all necessary development without being beholden to astronomy. It would be difficult to maintain that view today, as I hope to show in discussing some current examples of problems in astronomy which are as much, or even more, problems in physics, and which have widened our horizons as natural philosophers in a way that was not previously suspected.

Spectra of highly ionised atoms

I start with an example which shows that there is still interesting work to do even in the well-tilled field of atomic spectroscopy. When first rockets and subsequently artificial satellites were able to take spectrometers designed for wavelengths shorter than 3000 Å above the Earth's atmosphere, it was found that at wavelengths shorter than 2500 Å the continuum radiation from the Sun becomes very weak and the spectrum is dominated by emission lines which, it is now known, are radiated by the very hot tenuous gas in the corona. Many of the lines were quickly identified as coming from ionised states of such atoms as hydrogen, helium, carbon, oxygen, silicon and nitrogen, but a group of quite intense lines at wavelengths less than 200 Å (Plate 5) presented a problem. Members of the Astrophysical Research Unit at Culham suggested that they came from iron in very highly ionised states, and investigations in which the plasma source ZETA was used, established that that was indeed the explanation. You may perhaps wonder why there is an Astrophysical Research Unit at the Culham Laboratories, which were set up to investigate problems of power generation by fusion reactions. The answer is that physicists at Culham realised some while ago that the atmosphere of the Sun is a plasma source with which they could study some of the problems they were faced with in plasmas generated in laboratory instruments for the fusion programme. For this reason they were the first group in this country to make really precise spectroscopic observations of the solar chromosphere and corona at wavelengths shorter than 3000 Å, using instruments carried in rockets and satellites. I take the highly ionised lines as a striking example of collaboration between astronomy and physics, but ultra-violet astronomy is continually posing spectroscopic problems for theoretical and experimental study.

Galactic masers

Just about eight years ago, radio-astronomers at Berkeley observed radiations at 18 cm wavelength with such strange properties that they were hesitant to ascribe them to emission from the hydroxyl molecule, even though it was already known that hydroxyl molecules are present in clouds of inter-stellar gas and absorb 18 cm radio waves. It is now established that even though the aerial temperatures when looking at some sources reach 50 K, the sources are extremely small, and indeed not all can be resolved by interferometers with base lines extending right across the Earth, and in consequence the intensity of the radiation is equal to that which would be emitted by a black body at 10^{13} to 10^{15} K. The radiation is strongly polarised, and each tiny source,

of which there are usually many to be observed in an active region, is characterised by a particular frequency, corresponding no doubt to a particular velocity of gas along the line of sight, and by a sense of circular polarisation. The intensities of some sources vary significantly in weeks or months. These and other properties are incompatible with spontaneous emission of radiation from the hydroxyl molecule, as might occur in gas discharge lamps – instead, the emission by the hydroxyl molecules is stimulated by radiation falling on the molecule, just as in lasers (ter Haar and Giddy, 1973).

It now seems that hydroxyl maser sources are closely associated with regions of ionised hydrogen known as H-II regions, or with infra-red stars. Furthermore, hydroxyl sources are not the only maser sources, for water sources also show stimulated emission; they are not strongly polarised, but the intensities are as great and the variations in time are even more rapid.

A very great deal has been written on hydroxyl masers since they were first discovered, but I believe it is true to say that the major problems they present are still far from solution. Stimulated emission only exceeds absorption (which is also stimulated!) if the number of atoms or molecules in a state with higher energy exceeds the number in a state with lower energy, in violation of Boltzmann's rule for thermal equilibrium. Evidently we want to know how this inversion of population, as it is called, is brought

about. Many ideas have been proposed; all present difficulties and none has been investigated experimentally. We badly need experimental study of possible ways of inverting hydroxyl and water populations, but we also need to know more about what the processes have to accomplish. Four states of the hydroxyl molecules are involved and four transitions, at 1720, 1667, 1665 and 1612 MHz, can occur between them. The relative intensities of the transitions show that the relative populations of the four levels depend on conditions in the source; they may arise from the process of inversion, but in addition the process of stimulated emission may itself change the relative populations. Before we can be sure just what the processes of inversion have to accomplish, we need to know what stimulated emission may do to alter relative populations.

Interstellar masers thus pose problems in molecular physics as well as in astronomy, and they have shown that there are unexpected things to be discovered about even such well-studied molecules as hydroxyl and water.

My first two examples of current problems in astrophysics are substantial and interesting but, at the same time, they require only modest extensions to our existing knowledge of physics and one has already yielded to experimental study, and it is quite likely that the other will also. Because astrophysics poses many comparable problems, especially in atomic and molecular spectroscopy, a distinct

branch of physics, known as laboratory astrophysics, has grown up, stimulated particularly by the foundation of the Joint Institute of Laboratory Astrophysics in Boulder, Colorado, as a co-operative institute of the National Bureau of Standards and the University of Colorado. Laboratory astrophysics provides data for tackling problems in astrophysics where the principles are understood but the particular applications have to be worked out, and it also attempts to devise laboratory methods for studying novel problems of physics thrown up by astrophysics.

Pulsars

You will recall that the first pulsar was discovered here in Cambridge by Professor Hewish and his collaborators (Hewish, Bell, Pilkington, Scott and Collins, 1968); since then many examples have been found and the properties of pulsars have been thoroughly studied. Pulsars emit radio waves at regular intervals and the visible radiation from one also pulsates. It is now generally thought that the radiation comes from regions of plasma in the atmosphere of a star which rotates once in the interval (of the order of a second) between one pulse and the next.

Because the angular momentum of a star is conserved, the period of rotation of a pulsar will be proportional to the period of rotation of the star from which the pulsar collapsed, to the mass of the pulsar and to the square of its radius. The ratio of the speed of rotation of the pulsar to that of the original star is greater than a million times, so that the radius of the pulsar is less than a thousandth of that of a normal star and the density is much greater than a thousand million times normal stellar densities.

Einstein in 1925 applied the principle of quantum mechanics to the statistical theory of gases, and predicted that at a sufficiently high pressure a gas would become degenerate, meaning that many atoms would have the same energy. White dwarfs are stars which have used up most of their nuclear energy and have converted most of their hydrogen to helium, carbon and so on. The pressure of radiation is inadequate to keep them as large as stars still burning hydrogen and the pressure that supports them against their own attraction of gravity is that of a degenerate gas of electrons. They are therefore much smaller and much denser than other stars and they radiate only feebly. Because they are so condensed and thus have small moments of inertia compared to their uncollapsed forms, they must be rotating very fast to conserve angular momentum, but no unequivocal evidence has been obtained to support that inference.

Before the discovery of pulsars there had been speculations that a further stage of degeneracy might occur in which the density is so high that protons and electrons are bound as neutrons, which then form a degenerate solid or liquid of much greater density than the degenerate gas of a white dwarf.

The radius and moment of inertia would therefore be yet smaller and the speed of rotation greater. It is now generally accepted that pulsars are to be identified with degenerate neutron stars and, in consequence, there is considerable interest in trying to calculate in detail the structure and properties of those objects. The problems are mainly concerned with how very large numbers of particles behave when they exert strong forces on each other – internuclear forces in the case of neutron stars. Such problems occur in the physics of solids and liquids generally and we know that in degenerate systems large numbers of particles behave in a co-ordinated manner, for example pairs of electrons in super-conductors and helium atoms in the superfluid form. There has recently been a suggestion that helium-3 atoms might associate in pairs and move collectively as a superfluid. Now most liquids or gases become degenerate or solidify at higher temperatures as the pressure increases, and so, while degeneracy is observed only at low temperatures at laboratory pressures, it may be that at the very much higher pressures met in celestial bodies, not only the neutron fluid but other substances behave in a de-generate manner. In particular, it is tempting to speculate that hydrogen in the metallic form, as we think it is in Jupiter (Cook, 1972), may in the central part of the planet be under a high enough pressure to be a superfluid or superconductor. Jupiter has a magnetic field, the only other planet apart from the

Earth known to have one, and since we suppose that the Earth's field arises from a self-excited dynamo sustained by movements in the liquid conducting core of the Earth, it is natural to think that the same mechanism operates in Jupiter. That, however, supposes that the metallic hydrogen in Jupiter is liquid, but it might be solid, in which case a dynamo could not be maintained. Similar speculations are suggested by neutron stars. Is the material solid or liquid, may a neutron star have a solid core surrounded by a liquid shell, overlain by a thin crust of nuclear matter, and if so might that crust move in slabs, or plates, as we believe the crust of the Earth to do? Experiment at the high pressures in Jupiter, let alone neutron stars, is quite beyond us at present, and therefore the collective behaviour of matter at very high pressures poses severe and fundamental problems to those studying the theory of condensed matter.

Other problems

Apart from modest extensions of reasonably well-known atomic and molecular physics, I suggest there are four groups of phenomena where it is possible that new principles of physics may await discovery.

First, there is the behaviour of matter at very high pressures, as we observe it in neutron stars, white dwarf stars and the planets. The range of pressures is large and correspondingly we may study matter in which, successively, the elastic energy succeeds the energies characteristic of crystal, atomic, nuclear and sub-nuclear structure.

Secondly, there are phenomena involved in very high densities of energy, from which we may be able to observe new aspects of the interaction of radiation and matter. The optical spectra of objects with very large red shifts, properties of intense radio sources and features of the centres of galaxies may all contain clues.

Thirdly, the consequences of general relativity have been explored in very few ways so far by observation.

Fourthly, there may be very infrequent processes which can only be detected in the relatively empty regions of intergalactic space where more frequent events do not obscure them.

These are some ways in which current astronomy may stretch physics. I do not intend it to be a complete catalogue, for that would be to give an impression contrary to my purpose. I am arguing that recent observations in astronomy open new windows in physics – it is for us to go through them to explore the new possibilities, not to sit behind them and try to set limits.

CONCLUSION

I have indicated a number of observations in astronomy which are extending our ideas of physics. It is useful to distinguish two ways in which extensions occur, although, as with all such distinctions, the boundary is blurred. On the one hand, there are systems, in particular isolated atoms and molecules, the structure and behaviour of which we believe we understand in principle – we know what forces are involved and we know that atoms and molecules obey Schrödinger's equation – but the variety of possible situations is so great that we are unlikely to find all the interesting ones unless we are guided by observation; and, as I have indicated, the relative importance of various processes in atomic and molecular physics is quite different in astronomical situations from the usual laboratory situation, so that our attention is drawn to aspects of physics that otherwise we might not have noticed.

Astronomy also poses problems where our understanding of fundamentals is probably still incomplete, although we may still believe that they lie within the general scheme of terrestrial physics. The same ideas about the collective behaviour of condensed matter which have been developed to study problems of the solid state in the laboratory may well suffice to interpret astronomical observations of condensed bodies – namely those in which thermal radiation is unimportant.

But astronomical observation may also call for drastic revision of some ideas of physics. I think one should go very cautiously here – there are many instances from the early days of planetary theory onwards where new observations have been claimed to invalidate existing physics, but which have yielded to a more subtle analysis on accepted lines or a larger understanding of the inherent variety of solutions in classical dynamics or Schrödinger's equation or equations of transfer of radiation. Certain aspects of modern physical theory have indeed been only tentatively applied to astronomy. Consider the general theory of relativity, which was developed by Enistein largely independently of astronomical observation, but was then tested against observation in the three so-called classical tests of the deflection of light by the Sun, of the gravitational red shift and of the motion of the perihelion of Mercury. A key prediction of general relativity is the existence of the Schwarzschild radius, from within which radiation cannot escape, and now that we suspect that highly condensed neutron stars exist, it is natural to ask whether there may be even more condensed bodies which lie within their Schwarzschild radii. If so, how could we detect them? Incidentally, the name of 'black holes' given to such possible bodies goes back almost a century. Fr Angelo Secchi, who rebuilt and re-equipped the Roman College Observatory, and explicitly set himself to 'the physical study of celestial bodies', and was criticised by some who said he studied physics, not astronomy, wrote in 1877 'there is the interesting fact of the probable discovery of dark masses dispersed in space, whose existence will be revealed by the bright background of the sky on which they are projected. Until now these masses have been classified as "black holes", but this explanation is quite improbable...'

Beyond such extensions of ideas in which we have good reason to believe for other reasons, there remains the possibility that astronomy will demand some more radical revision of ideas of physics, comparable with the development of quantum mechanics. The forces involved in continuous creation would have been of such a sort, and more recently Sir Fred Hoyle has been proposing that we need a new approach to the problem of how particles interact at great distance.

But let us return to Earth! I do not want to leave you with the impression that there is now nothing worth doing in the physics laboratory and that every physicist should forthwith become an astronomer, nor on the other hand do I countenance the idea that astronomy is just applied physics, for one of my points has, after all, been the interaction of astronomy and physics, that astronomical observations suggest theoretical or experimental lines on which physicists should work in order to elucidate problems in the more controllable circumstances of the laboratory, while physics guides us to an understanding of astronomical objects and processes. In any case there are major fields of physics, especially high-energy physics and the physics of condensed matter, which so far have owed little to astronomy, and as for astronomy, it is to some extent the study of particular objects or systems – stars, galaxies, the Universe – and not of general laws. So, as a natural philosopher by title, I would like to conclude with some ideas of the scope for laboratory physicists.

To begin with, all the astronomical observations which I have discussed came to be made because of the development of a new technique of observation – Wollaston's spectroscope and Huggins's use of the dry photographic plate in the past, radio-astronomy and astronomy from rockets and artificial satellites today. There will always be scope for physicists to devise new methods of observation.

The second field is the exploration of what I may call 'understood' physics, where we believe we are sure of the principles but where the variety of possible solutions of Laplace's equation, Schrödinger's equation, and so on, still provides us with many problems which are both intellectually challenging and of practical importance. Professor Pippard in his own Inaugural Lecture nearly twenty months ago (Pippard, 1972) was concerned to draw attention to the gap between understanding the mechanical principles of various problems and being able to work out in detail how a system would behave. We cannot possibly hope to list all possible solutions, even of Laplace's equation – we need the guidance of experiment and observation to choose those of interest and importance, and astronomy provides us with ideas. In contrast, I do not think that astronomy has much to offer in the way of new ideas to the high-energy physicist – it seems to me that the control of conditions, the observation of the interaction of otherwise isolated particles, are necessary at this stage when the theoretical framework to use has still to be constructed.

Laboratory astrophysics is now a major field of study; it supports astronomy by checking ideas and supplying data, especially in atomic and molecular physics, in which there has been a striking resurgence in the last fifteen years. Most laboratory astrophysics is concerned with atomic or molecular spectroscopy in the range of wavelengths from a few tens of nanometres to hundreds of micrometres, but with the recent discoveries of molecules in interstellar space and of X-ray and γ-ray sources, much more work will be needed on spectroscopy in the microwave, X-ray and γ-ray ranges of energy. There are major problems here alike in theory and experiment and, as the history of the gas laser shows, laboratory astrophysics may have large consequences in engineering.

Finally there are what I may call 'way-out' experiments, experiments in which we stretch our technical capabilities to the limit to try to grasp some fundamental aspect of nature – I think of Dicke's work on the constant of gravitation, or that of Weber on gravitational radiation. How can we bring eternity into a grain of sand, the vastness of space on to our little Earth, minute forces into competition with the disturbances of our laboratories? Here are the real challenges to deep thinking and ingenious experiment, which few have the temerity to accept.

I like to predict that, save in high-energy physics, major advances in natural philosophy will develop from ideas suggested by astronomical observations. I am inclined, however, to think that this idea may be better as a guide to conduct than as an insight into the future – tomorrow it may well be proved wrong.

I am indebted to Dr D. W. Dewhirst for obtaining for me the material for Figure 2 and Plates 1–4.

REFERENCES

BALMER, J. J. (1885). Notiz über die Spectrallinien der Wasserstoffs. *Verh. d. Naturf. Gesel. Basel* **7**, 548–60.

CHARLTON, T. M. (1972). The development of engineering education. *Aberdeen Univ. Rev.* **44**, 331–8.

COOK, A. H. (1972). The dynamical properties and internal structures of the Earth, the Moon and the Planets. *Proc. Roy. Soc. A.* **328**, 301–36.

HEWISH, A., BELL, S. J., PILKINGTON, J. D. H., SCOTT, P. F. and COLLINS, R. A. (1968). Observations of a rapidly pulsating radio source. *Nature (Lond.)* **217**, 709–13.

HUGGINS, W. (1880). On the photographic spectra of stars. *Philos. Trans. Roy. Soc.* **171**, 669–90.

KIRCHHOFF, G. (1862). *Researches on the Solar Spectrum and the Spectra of the Chemical Elements.* Transl. H. E. Roscoe. 36 pp. London and Cambridge: Macmillan.

KIRCHHOFF, G. (1863). *Researches on the Solar Spectrum and the Spectra of the Chemical Elements.* Transl. H. E. Roscoe. Second Part. 16 pp. Cambridge and London: Macmillan.

LIVEING, G. D. and DEWAR, SIR J. (1915). *Collected Papers on Spectroscopy.* xv + 566 pp. Cambridge University Press.

LOCKYER, J. N. (1874). *Contributions to Solar Physics.* xxi + 676 pp. London: Macmillan.

PEVSNER, N. (1970). *Robert Willis.* 27 + unnumbered pp. Northampton, Mass.: Smith College Studies in History.

PIPPARD, A. B. (1972). *Reconciling Physics with Reality.* 40 pp. Cambridge University Press.

TER HAAR, D. and GIDDY, M. A. (1973). Interstellar hydroxyl water and formaldehyde masers. *Rep. prog. phys.* (in press).

WILLIS, R. (1841). *Principles of Mechanism.* xxxi + 446 pp. London: J. W. Parker.

WILLIS, R. and CLARK, J. W. (1886). *The Architectural History of the University of Cambridge and of the Colleges of Cambridge and Eton.* 4 vols. Cambridge University Press.

This Inaugural Lecture was delivered on
23 May 1973